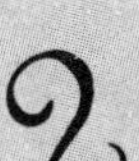

MENTAL MATHS

Part-5

Pooja Yadav
(M.A. B.ED)

Published by:

THE BOOK PARADISE

Head Office: 13/11, Subhash Nagar,

Old Railway Road, Gurgaon-122001, (India)

Phone: 0124-4062330, 4067543 M: 9871249090

E-mail: bookparadise@hotmail.com

Show Room: 4598/12-B, Ansari Road, Darayaganj, New Delhi-2

First Published in 2009 by

THE BOOK PARADISE

13/11, Subhash Nagar,

Old Railway Road, Gurgaon-122001, (India)

Reprint Edition - 2014

Designed by:

Kingston Koncepts

Printed at:

Print India Press

Find the sum :-

$$\begin{array}{r} 167234 \\ 2932 \\ +\ 709657 \\ \hline \\ \hline \end{array}$$

$$\begin{array}{r} 93210072 \\ 47210029 \\ +\ 872999 \\ \hline \\ \hline \end{array}$$

Find the difference :-

$$\begin{array}{r} 7525650 \\ -\ 2939999 \\ \hline \\ \hline \end{array}$$

$$\begin{array}{r} 7650789 \\ -\ 550759 \\ \hline \\ \hline \end{array}$$

Find the product :-

$$\begin{array}{r} 52236 \\ \times\ 273 \\ \hline \\ \hline \end{array}$$

$$\begin{array}{r} 495890 \\ \times\ 789 \\ \hline \\ \hline \end{array}$$

Divide

$35\overline{)278469}$ $25\overline{)795633}$

Solve the following:-

80,000 × 10 =

1,16,100 × 100 =

53, 85, 100 ÷ 1000 =

7,000 × 20 =

50,000 × 30 =

Rewrite the numeral marking the periods

1. 44379862 ____________________

2. 95072525 ____________________

3. 70707710 ____________________

4. 5525025 ____________________

Write the numeral for

1. Twenty crore, sixty three lakhs, five hundred thirty nine -

   ____

2. Twelve crore, fifty lakhs, nine thousand nine -

   ____

3. Forty seven crore, six lakh, five hundred six -

   ____

4. Five crore five -

   ____

Find the product

$\frac{8}{7} \times 7 = \frac{8}{\cancel{7}} \times \cancel{7} = \frac{8}{1} \times \frac{1}{1} = \frac{8}{1} = 8$

$\frac{2}{3} \times \frac{3}{6} \times \frac{5}{2} =$ ____________ = ____________ = ____________

$\frac{1}{2} \times \frac{1}{5} \times \frac{1}{4} =$ ____________ = ____________ = ____________

$\frac{18}{25} \times 100 =$ ____________ = ____________ = ____________

Find in the blanks

$2\frac{1}{6} \times 3\frac{5}{3} = 3\frac{5}{3} \times$ ________

$\frac{1}{7} \times 7 =$ ________

$\frac{2}{9} \times \frac{9}{2} =$ ________

$\frac{5}{7} \times$ ________ $= 1$

Shade the given fraction as indicated

$\frac{2}{4}$

$\frac{1}{2}$

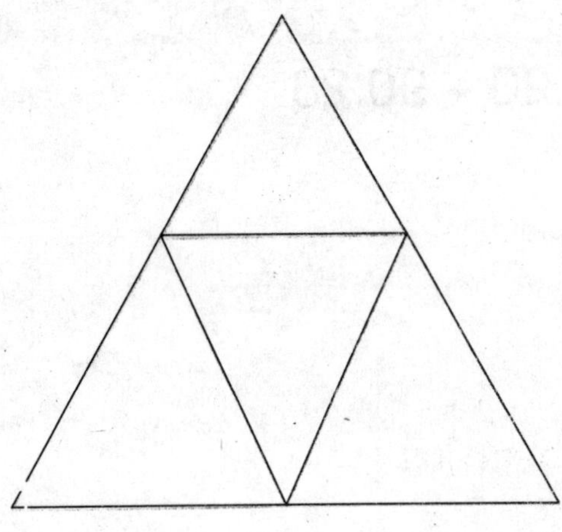

$\frac{1}{3}$

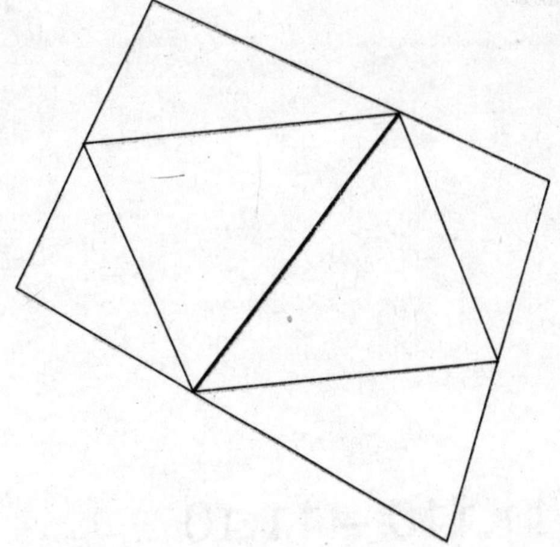

$\frac{1}{3}$

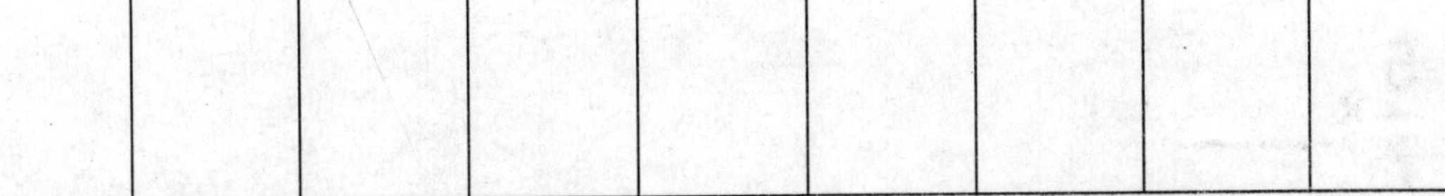

Find the sum of :

1. 715.03 + 85.29

2. 39.23 + 77.23 + 90.23

3. 94.29 + 79.897

4. 111.008 + 111.110 + 11.10

Find the difference of

1. 125.285 and 111.095

2. 993.25 and 773.75

3. 94.29 and 78.99

4. 4.24 and 6

Fill in the blanks

1. 6.628×100 = ______________

2. 8.7×10 = ______________

3. 0.008×10 = ______________

4. 0.05×1000 = ______________

5. 19.23×10 = ______________

6. 29.10×1000 = ______________

Find the product

1. 6.99×4 = ______________

2. 25.10×10 = ______________

3. 1.12×2.2 = ______________

4. 0.05×2.6 = ______________

Convert in decimals

$2\frac{7}{30} =$ ______________________________

$9\frac{9}{10} =$ ______________________________

$7\frac{5}{20} =$ ______________________________

$2\frac{7}{50} =$ ______________________________

$9\frac{1}{10} =$ ______________________________

$7\frac{7}{20} =$ ______________________________

Divide :

1. 13.03 by 3

2. 24.450 by 9

3. 260.33 by 30

4. 35.39 by 5

Solve :-

1. $29 - 11 + 2 =$

2. $2.2 \times 2.02 - 2 + 201 =$

3. $7 - 75 \div 15 + 3 - 1.5 =$

4. $3.2 \times 2.3 + 2 - 3.2 =$

Put ✓ or ×

$\frac{5}{2} \times 1 = \frac{1}{2} \times 5$ ☐

$7\frac{5}{6} = 7\frac{6}{5}$ ☐

$\frac{1}{2} \times 3 = \frac{3}{2} \times 1$ ☐

$2\frac{4}{7} \times 1 = 2\frac{4}{7}$ ☐

Find the product of

1. $\frac{8}{7}\times7=$ ________________

2. $3\frac{1}{2}\times5=$ ________________

3. $6\frac{1}{3}\times8=$ ________________

4. $\frac{18}{25}\times100=$ ________________

5. $\frac{1}{5}\times\frac{1}{4}=$ ________________

6. $\frac{2}{3}\times\frac{3}{2}\times\frac{5}{2}=$ ________________

7. $\frac{6}{5}\times\frac{5}{3}=$ ________________

8. $\frac{3}{12}\times4=$ ________________

Fill in the blanks

1. $2\frac{1}{7} \times 3\frac{3}{5} = 3\frac{3}{5} \times$ ______

2. $\frac{7}{13} \times \frac{13}{7} =$ ______

3. $\frac{8}{5} \times 1\frac{3}{5} =$ ______

4. $12\frac{1}{4} \times \frac{40}{4} =$ ______

5. $\frac{1}{12} \times 12 =$ ______

6. $\frac{5}{13} \times$ ______ $=1$

7. $\frac{7}{17} \times$ ______ $=0$

8. $\frac{2}{5} \times \frac{5}{2} =$ ______

Solve the following

1. $\frac{8}{3} \div \frac{2}{6} =$ ______

2. $0 \div 1\frac{1}{7} =$ ______

3. $7\frac{1}{2} \div 6\frac{1}{2} =$ ______

4. $81 \div \frac{9}{3} =$ ______

5. $110 \div \frac{11}{9} =$ ______

6. $12\frac{1}{4} \div \frac{7}{5} =$ ______

7. $70 \div \frac{7}{12} =$ ______

8. $8\frac{1}{2} \div 2\frac{1}{2} =$ ______

Solve

1. A cloth of length $12\frac{1}{2}$ m has to be divided equally into three pieces. What will be the length of each piece?

2. The height of a building is $10\frac{1}{2}$ m what will be the height of 5 such buildings ?

3. A sack contains Kg. $25\frac{1}{2}$ of rice. What will be the weight of 10 such sacks?

4. 10m of ribbon has to be cut into the pieces of $\frac{1}{2}$ m long. What will be the number of pieces of ribbon?

5. A truck contains $75\frac{1}{2}$ Kg. of fruit. How much fruit can be carried into 8 such trucks?

Convert into fractions

1. 0.230 =

2. 510.6 =

3. 29.12 =

4. 0.660 =

5. 0.04 =

Write decimals for the following

1. 4 tens + 6 ones + 2 tenth + 3 hundredths =

2. 6 hundredths + 3 thousandths =

3. 2 hundreds + 1 ones + 7 thousandths =

4. 1 tens + 3ones + 1 tenths + 3 hundredths =

Put > or <

1. 12.89 ☐ 18.29

2. 0.605 ☐ 0.625

3. 300.300 ☐ 30.300

4. 110.80 ☐ 111.80

5. 87.085 ☐ 87.875

6. 125.11 ☐ 125.15

7. 460.20 ☐ 406.02

8. 111.60 ☐ 110.16

Solve:

1. The cost of 6 toys is Rs. 126.60. What is the cost of 1 toy?

2. 1 Lt milk costs Rs. 23.50. What will be the cost of 20 Lt. of milk.

3. The capacity of a tanker is 500 Lt. It is filled 250.95 Lt. How much more water will be filled to fill it completely?

4. 20 pencils costs Rs. 105.50. What will be the cost of 1 pencil?

5. A man purchased T.V. for Rs. 11065.50, a DVD for Rs. 4625.75 and a fridge of Rs. 9760.75. What is the total amount he will have to pay?

Put ✓ or ×

1. Selling price = cost price + gain

2. Cost price = selling price + gain

3. Selling price = cost price loss

4. Gain + cost price = selling price

Fill in the blanks :

1. If S.P. and C.P. are same there is no ______ or ______.

2. If S.P. is more than C.P. there is ________.

3. If C.P. is more than S.P. there is ________.

4. Gain or loss is calculated on __________.

Fill in the blanks

1. A shopkeeper sold an item for Rs. 2500 and lost Rs. 300. The cost price of the item was _________.

2. 10 Kg. of fruit was bought for Rs. 500 and sold for Rs. 700. The profit / loss was _________.

3. 3 dozen of eggs were sold for Rs. 100 at a loss of Rs. 25. The cost price was _________.

4. Cost price of a gown is Rs. 160 and selling price is 175. The profit is _________.

5. A man bought a T.V. for Rs. 10,050 and sold it with a loss of Rs. 100. The S.P. is _________.

Complete the table

Selling price	Cost price	Profit	Loss
9500	______	100	______
1700	950	______	______
______	2300	400	______
700	350	______	______
290	275	______	______
16650	18000	______	______
7850	8250	______	______
______	6200	750	______

Complete the table

Figure	Length (m)	Breadth (m)	Area (Sq. m.)
Rectangle	5	3	______
Square	10	______	100
Rectangle	8	______	32
Rectangle	91	19	______
Square	18	18	______
Rectangle	17	______	51
Rectangle	______	4	36
Square	______	______	81

Find the area

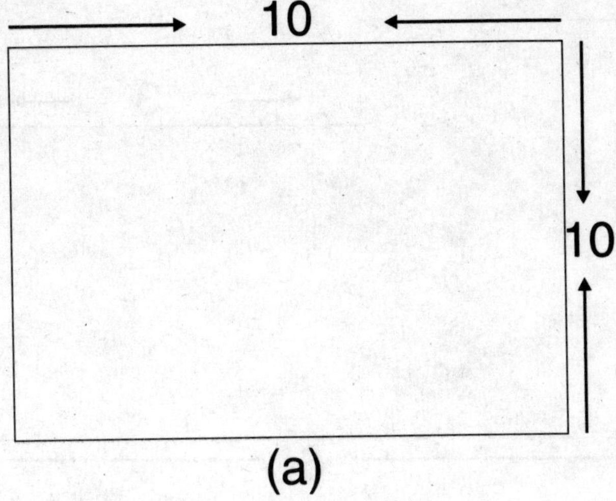

(a)

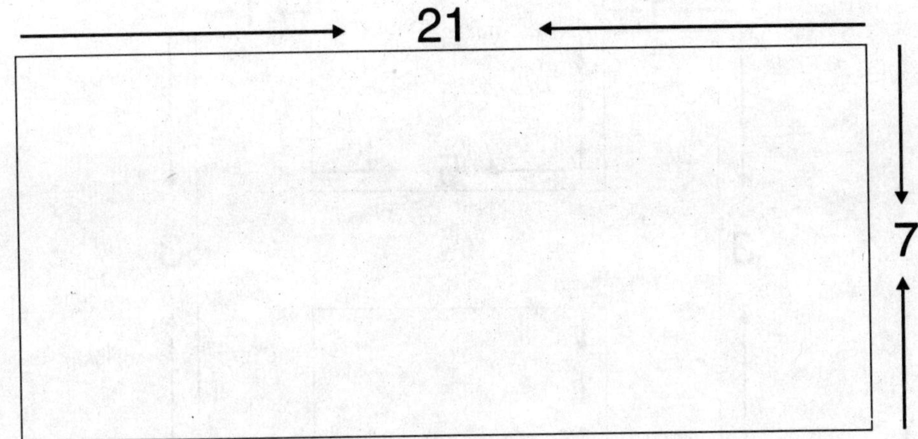

(b)

Find the area

(a)

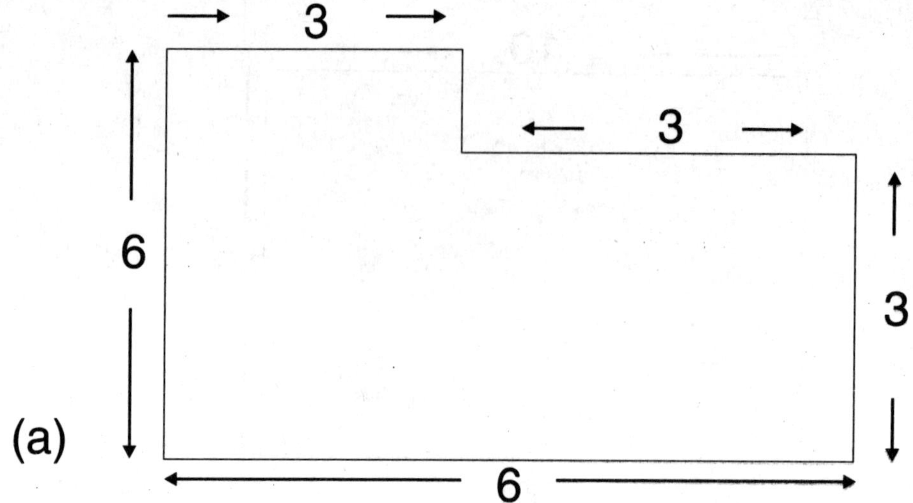

(b)

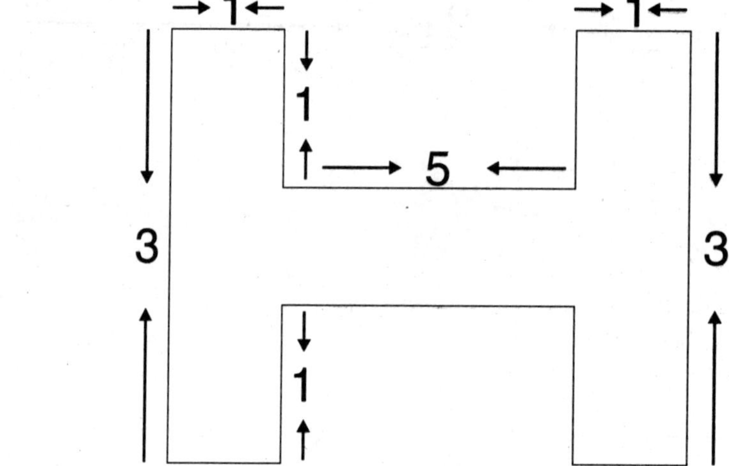

Fill in the blanks

1. The area of a square is 121 square cm. The length of its side is ______.

2. The length of a rectangle is 4m. 25cm. and its breadth is 2m. 10 cm. Its area is _____

3. The area of a square with a side 15 is ______

4. The width of a rectangle will be _______ if its length is 144 m. and area is 14400 sqm.

5. The side of a square will be ______ if its area is 100000 Sq.cm.

Express each of the following as decimals

1. $\frac{17}{100}$ = ____

2. $\frac{3}{10}$ = ____

3. $\frac{111}{100}$ = ____

4. $\frac{15}{10}$ = ____

Reduce to the lowest terms

1. $\frac{95}{100}$ = ____

2. $\frac{480}{10}$ = ____

3. $\frac{500}{100}$ = ____

4. $\frac{500}{50}$ = ____

Write as a percentage

2.15 = ____________________

6.3 = ____________________

9.07 = ____________________

0.51 = ____________________

0.06 = ____________________

7.17 = ____________________

3.05 = ____________________

6.6 = ____________________

0.12 = ____________________

1.8 = ____________________

Convert into fraction with lowest term and decimals.

1. 82% = ____________________, ____________________

2. 115% = ____________________, ____________________

3. 91% = ____________________, ____________________

4. 37% = ____________________, ____________________

5. 42% = ____________________, ____________________

6. 110% = ____________________, ____________________

7. 75% = ____________________, ____________________

8. 92% = ____________________, ____________________

9. 15% = ____________________, ____________________

10. 99% = ____________________, ____________________

Change into percentage

1. $3\frac{2}{4}$ = ______________________

2. $6\frac{9}{2}$ = ______________________

3. $4\frac{2}{10}$ = ______________________

4. $7\frac{3}{10}$ = ______________________

5. $\frac{15}{4}$ = ______________________

6. $\frac{25}{2}$ = ______________________

7. $\frac{2}{10}$ = ______________________

8. $2\frac{3}{5}$ = ______________________

Circle the correct equivalent

0.06 → $\frac{60}{100}$ $\frac{6}{100}$ $\frac{600}{1000}$

5% → 5, .5, .05

9.01 → 901%, 90.1%, 9.01%

115% → 1.15, 11.5, 0.115

What percentage of a rupee is

70 paise = ____________

65 paise = ____________

40 paise = ____________

20 paise = ____________

28 paise = ____________

50 paise = ____________

Fill in the blanks

15% of 220 = ____________

20% of 680 = ____________

19% of 200 = ____________

65% of 325 = ____________

26% of 900 = ____________

10% of 2300 = ____________

99% of 65200 = ____________

72% of 1000 = ____________

Put >, < or =

15% of 100		20 % of 200
25% of 1000		50% of 100
65% of 200		20% of 100
50% of 150		30% of 200
30% of 500		30% of 600
8% of 100		5% of 500
25% of 750		20% of 1000
5% of 1000		10% of 2000

Find the value of

70% of 1000 lt = ____________

35 % of 1000 Rs. = ____________

9% of 720 grams = ____________

20% of 100 mm = ____________

$12\frac{2}{3}$% of 100m = ____________

5% of 100 Rs. = ____________

75% of 200lt. = ____________

90% of 150 Rs. = ____________

What % of

1. 90 litres is of 150 Lt.

2. 25 m. is of 100 m.

3. 200 Rs. is of 1000 Rs.

4. 250 g. is of 1000 g.

5. 150 Kg. is of 1000 Kg.

Fill in the blanks

1. C.P. = Rs. 2000

   Gain % = 6%

   S.P. = ____________

2. C.P. = Rs. 400

   S.P. = Rs. 800

   Gain % = ____________

3. C.P. = Rs. 1000

   S.P. = ____________

   Loss % = 5%

4. C.P. = Rs. 800

   S.P. = Rs. 500

   Loss % = ____________

Solve :-

1. By selling a shirt for Rs. 500, a shopkeeper looses 20%. Find its cost price.

2. The cost price of a T.V. is Rs. 20,000. The shopkeeper sells it at a profit of 10%. What will be the selling price?

3. Mr. Sharma purchases a car for Rs. 110900. He spent Rs. 1100 for its repairing. He sold it at a profit of 5%. What is the selling price of the car?

4. A man purchases a house for Rs. 111000 and spent Rs. 5500 for its repairing. He sold it for Rs. 139800. What is his gain or loss %?

5. The C.P. of an item is Rs. 360. The overhead charges are Rs. 50. The loss is 6%. What is the S.P.?

Complete the table :-

Remember $Speed = \frac{Distance}{Time}$

Speed	**Distance**	**Time**
________	125 km.	5 hours
80 km/hr.	800 km.	________
________	1500 km.	1 hr. 50 min.
68.75 km/h.	________	4 hours
18.75 km/h.	________	8 minutes
33.3 km/h.	600 km.	________

Fill in the blanks :-

1. 1 m. = ______________ Km.

   So, 20 m = ______________ Km.

2. 1 Second = ______________ hours

   So, 25 m/s = ______________

   = ______________ Km/hour

3. 36 Km/h = ______________ m/s

4. 2000 m/s = ______________ km/h

Put > or <

1. 75 Km/h ○ 15m/s
2. 25 Km/h ○ 20m/s
3. 20 Km/h ○ 20 m/s
4. 30 Km/h ○ 45 m/s

Solve :-

1. A car covers a distance of 40 Km. 20 m. in 4 hours. What is its speed?

2. A man travels 4 km. in 20 min. What is his speed?

3. A train runs at a speed of 120 Km/h and covers a distance in 4 hours. What is the distance?

4. With a speed of 80 Km/h in how much time will a bus cover 320 Km?

5. A man has to cover a distance of 300 Km. in 3 hours. What should be his speed?

Prepare a bill for the following items :-

8 notebooks for Rs. 10 each.

20 erasers for Rs. 2 each.

3 chart papers for Rs. 5 each.

5 colour boxes for Rs. 40 each.

10 colour pencils for Rs. 8 each.

S.No.	Item	Price/Item	Qty.	Amount

Test Exercise 1

Solve the following :-

1. $7000 \div 7$ = ____________________

2. $18920000 \div 1000$ = ____________________

3. 7000×5000 = ____________________

4. 150000×100 = ____________________

5. 9000×50 = ____________________

Fill in the blanks :-

	3	_	_	2
+	5	2	1	_
	8	7	9	7

	_	3	2	2
	3	5	_	2
+	3	5	7	7
	9	_	5	_

Test Exercise 2

Find the difference and check the result.

1. 23509 – 9009

2. 29000 – 300

3. 267529 – 2900

4. 10000 – 200

Divide and check

$21\overline{)650293}$

$21\overline{)650293}$

Test Exercise 3

1. $\frac{20}{25} \times 100 =$ ________

2. $\frac{63}{15} \times 25 =$ ________

3. $\frac{2}{5} \times \frac{1}{2} \times \frac{1}{2} =$ ________

4. $\frac{2}{3} \times \frac{3}{2} =$ ________

Arrange in increasing order :-

1. 6.3, 6.03, 60.3, 6.003

____

2. 11.11, 11.01, 11.10, 1.11

____

3. 25.5, 2.55, 25.05, 25.50

____

4. 10.5, 1.05, 1.50, 15.0

____

Express them into fractions with lowest terms and decimals.

75% = ____________, ____________,

120% = ____________, ____________,

88% = ____________, ____________,

300% = ____________, ____________,

Change them into percentages

2.55 =

6.10 =

$6\frac{3}{4}$ =

$7\frac{7}{10}$ =

Find the area

1.

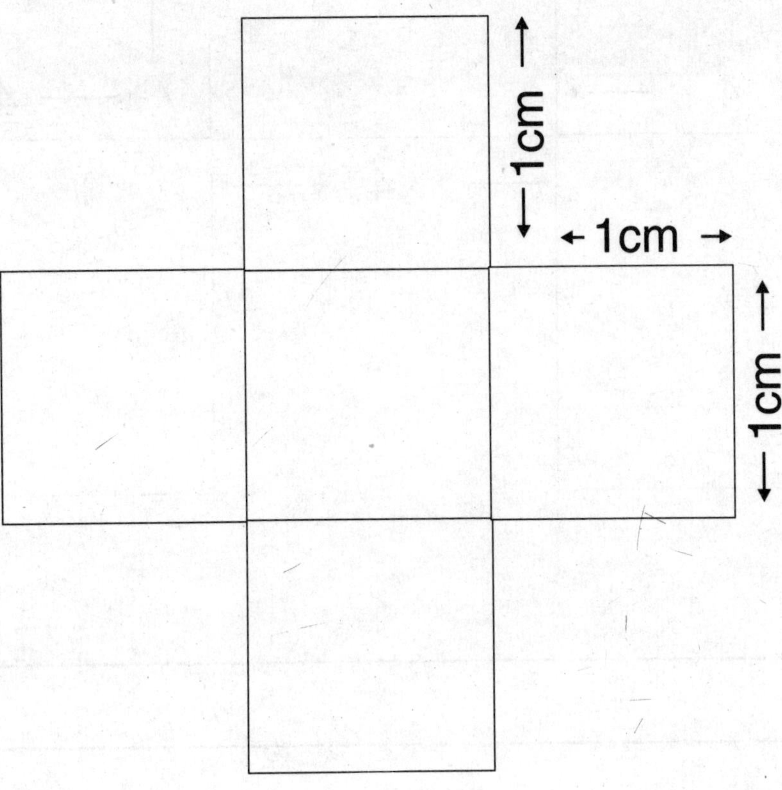

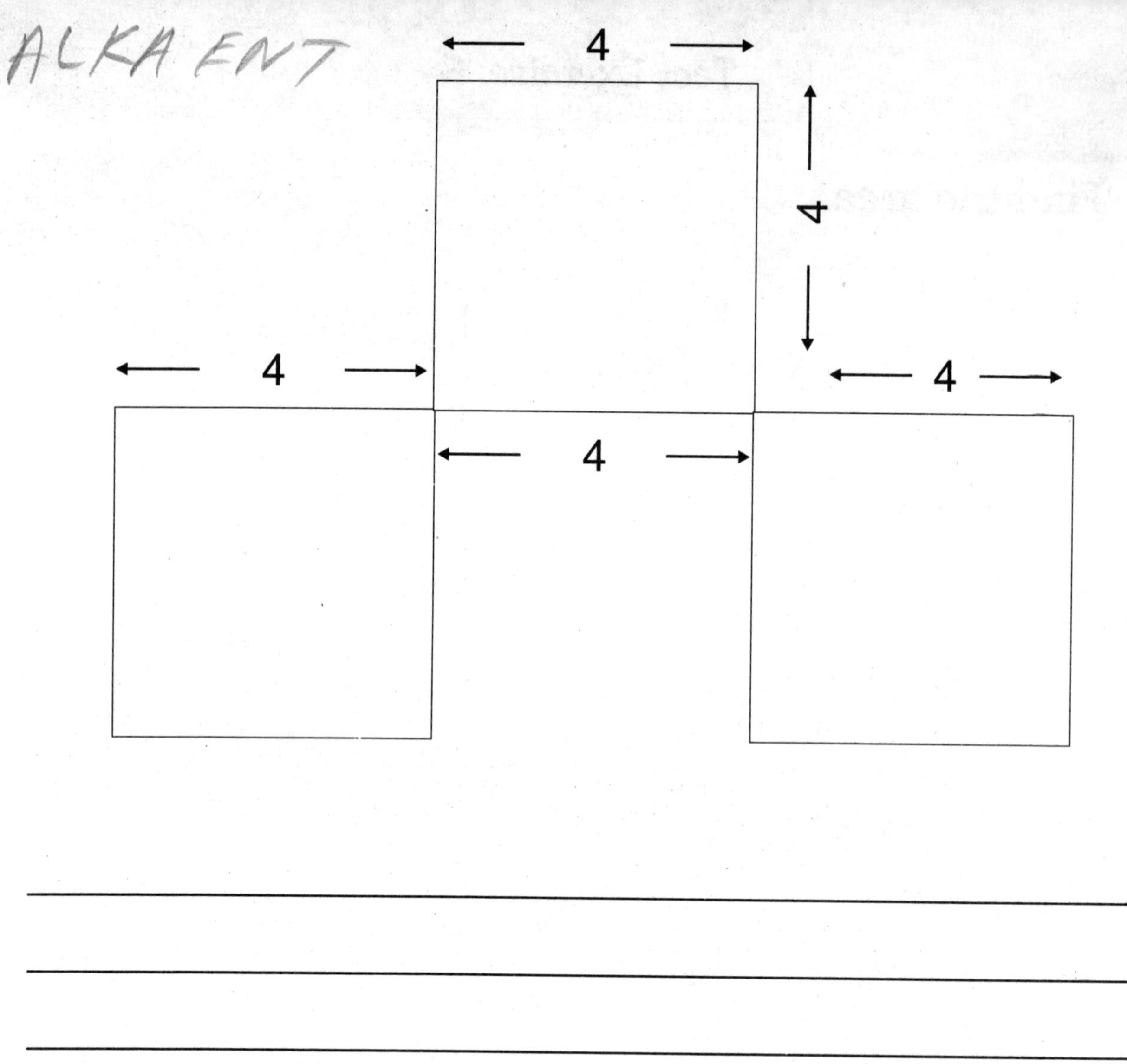

Solve-

1. A person sold 60% of the 250 tickets. The number of the tickets sold was ______.

2. A person sold a fridge for Rs. 10,000 with a profit of 10%. What is the C.P. of the fridge?